Louis Pasteur : the vaccine inventor

René Vallery-Radot
John Tyndall

Louis Pasteur
The Vaccine Inventor

LM Publishers

Louis Pastor was a distinguished French chemist and author of researches in microbial fermentation. He is considered as one of the great scientists of the world mainly for his remarkable breakthroughs in the causes and preventions of diseases.

LOUIS PASTEUR.

The Life-Work of Louis Pasteur[1]

Louis Pastor passed his childhood in a small tannery which his father had bought in the city of Arbois, in the department of the Jura, to which he removed from the ancient city of Dôle, in the same department, where he was born. When Louis became of suitable age, he was sent to the communal school, and was so proud of the fact that, though he was the smallest of the pupils, he went on the first day with his arms full of dictionaries away beyond his years. He does not appear, as yet, to have been a particularly diligent student. He was as likely to be found drawing a portrait or a sketch—and the walls of several Arboisian

---

[1] René Vallery-Radot

houses bear testimonies of his skill in this art—as studying his lesson, and to go a-hunting or a-fishing as to take the direct way to the school. Yet the principal of the college was ready to predict that it was no small school like this one, but some great royal institution, that was destined to enjoy his services as a professor. As there was no Professor of Philosophy in the college at Arbois, young Pasteur went to Besançon to continue his studies. Here, in the chemistry-class, he so vexed Professor Darlay with his frequent and searching questions, that the old gentleman was disconcerted, and declared it was his business to question the pupil, not Pasteur's to question him. Pasteur then had recourse to a pharmacist in the town who had gained some distinction in science, and took private lessons in chemistry from him. He fared better at the *École Normale*, where he had Balard for a teacher, and also enjoyed the instructions of Dumas, with whom he formed a life-long friendship at the Sorbonne.

Pasteur's first important investigation was suggested at about this time, by an observation of Mitscherlich, the German mineralogist, of a difference in the behavior toward polarized light of the crystals of paratartrate of soda and ammonia and tartrate of soda and ammonia, bodies identical in composition and external form and other properties. Pasteur discovered differences in the form of the crystals and the molecular structure of the two bodies that had escaped detection, and was led to consider that all things may be divided into two categories: those having a plane of symmetry—that is, capable of being divided so that the parts on either side of the plane of division shall be equal and identical—or symmetrical bodies; and dissymmetrical bodies, or those not capable of being so divided. Occupied with the idea that symmetry or dissymmetry in the molecular arrangement of any chemical substance must be manifested in all its properties capable of showing the quality, he pursued his investigations till he reached the conclusion that an essential difference in properties as to symmetry exists between mineral and dead matter and matter in which life is in course of development, the former being symmetrical, the latter unsymmetrical.

Pasteur's wedding-day came on while he was engaged in this investigation. He went, not to the marriage-feast, but to his laboratory, and had to be sent for when all was ready.

With his observing powers quickened by his studies of symmetry and dissymmetry, Pasteur went to the researches with which his life has been identified, beginning with his studies in fermentation. Liebig's theory, that fermentation is a change undergone by nitrogenous substances under the influence of the oxygen of the air, ruled at the time, and the observations of Schwann and Cagniard-Latour on the yeast-plant were overlooked or regarded as exceptional. M. Pasteur continued the investigation of the alcohol-producing yeast-plant, and, cultivating it in suitable solutions, proved that it possessed organizing power ample to account for the phenomena. He found a similar organism—minute cells or articulations narrowly contracted in the middle—active in the lactic fermentation, capable of cultivation; and another organism, a vibrion, full of motion, living singly or in chains, working in the butyric fermentation.

The butyric vibrion was found to work quite as vigorously and with as much effect when no air was added to the decoctions, and in fact to perish with a stoppage of the formation of butyric acid when air was too freely supplied.

Reverting to the development of the yeast-plant and the alcoholic fermentation, he found that they also went on best when free air was excluded. Thus, Liebig's dictum, that fermentation is the result of the action of oxygen, must be reversed or abandoned. The organisms working these processes were given the class-name of *anærobes* or beings that live without air. The French Academy's impressions of the results of Pasteur's work were spoken by Dumas, who said to him, "In the infinitely little of life you have discovered a third kingdom to which belong those beings which, with all the prerogatives of animal life, have no need of air to live, and find the heat they require in the chemical decompositions they provoke around them." The place of the organisms in the economy of Nature had not yet been fixed, but Pasteur was able to declare: "Whether the progress of science makes the vibrion a plant or an animal, is no matter; it is a living being endowed with motion, that lives without air and is ferment." It would be mere repetition to follow the experiments in putrefaction, where Liebig had denied that living organisms have any place, into which Pasteur carried the same methods and obtained the same results as in the case of fermentation. He proved that living organisms have all to do with it.

After M. Pasteur had been collecting his proofs for twenty years, Dr. Bouillaud sharply asked in the Academy: "How are your microscopic organisms disposed of? What are the ferments of the ferments?" He, as well as Liebig, believed the question could not be answered. Pasteur proved, by a series of the parallel experiments of the kind that have since become familiar, that oxygen deprived of its germs is incapable of producing fermentation or putrefaction, even after years, while the same substances are acted upon at once if the germs are present; and then answered that the ferments are destroyed by a new series of organisms—*ærobes*—living in the air, and these by other aerobes in succession, until the ultimate products are oxidized. "Thus, in the destruction of what has lived, all is reduced to the simultaneous action of the three great natural phenomena — fermentation, putrefaction, and slow combustion. A living being, animal or plant, or the *débris* of either, having just died, is exposed to the air. The life that has abandoned it is succeeded by life under other forms. In the superficial parts accessible to the air, the germs of the infinitely little ærobes flourish and multiply. The carbon, hydrogen, and nitrogen of the organic matter are transformed, by the oxygen of the air and under the vital activity of the ærobes, into carbonic acid, the vapor of water, and ammonia. The combustion continues as long as organic matter and air are present together. At the same time the superficial combustion is going on, fermentation and putrefaction are performing their work, in the midst of the mass, by means of the developed germs of the anaerobes, which not only do not need oxygen to live, but which oxygen causes to perish. Gradually the phenomena of destruction are at last accomplished through the work of latent fermentation and slow combustion. Whatever animal or vegetable matter is in the open air or under the ground, which is always more or less impregnated with air, finally disappears. The processes can be stopped only under an extremely low temperature, . . . in which the microscopic organisms cannot flourish. These facts come in to fortify the still new ideas of the part which the infinitely little play as masters of the world. If their work, always latent, were suppressed, the surface of the globe, overloaded with organic matters, would become uninhabitable."

Pasteur extended his observations to the acetic fermentation, or conversion of alcohol into vinegar, in which he found an organism, the *Mycoderma aceti*, actively promoting a process of oxidation. Liebig had attributed this fermentation, also, to the presence of an albuminoid body in process of alteration, and capable of fixing oxygen. He knew of the plant called "mother," but regarded it as an outgrowth of the fermentation, and in no sense the cause. Pasteur proved, by experiments that left no room for doubt—the prominent characteristic feature in all his investigations—that the plant is the real agent in producing the fermentation. He eliminated from his compositions the albuminoid matter, which Liebig had declared to be the active agent, and replaced it with crystallizable salts, alkaline phosphates, and earths; then, having added alcoholized water, slightly acidulated with acetic acid, he saw the mycoderm develop, and the alcohol change into vinegar. Having tried his experiments in the vinegar-factories at Orleans, he became so sure of his position that he offered to the Academy, in one of its discussions, to cover with the mycoderm, within twenty-four hours, from a few hardly-visible sowings, a surface of vinous liquid as extensive as the hall in which they were meeting.

Liebig allowed ten years to pass after Pasteur's investigations, and then published a long memoir traversing his conclusions. Pasteur visited Liebig at Munich, in 1870, to discuss the matter with him. The German chemist received him courteously, but excused himself from the discussion, on the ground of a recent illness. The Franco-German War came on; but, as soon as it was over, Pasteur invited Liebig to choose a committee of the Academy, and furnish a sugared mineral liquid. He would produce in it, before them all, an alcoholic fermentation in such a way as to establish his own theory and contradict Liebig's. Liebig had referred to the process of preparing vinegar by passing diluted alcohol through wooden chips, as one in which no trace of a mycoderm could be found, but in which the chips appeared perfectly clean after each operation. It was, in fact, impossible that there should be any mycoderm, because there was nothing on which it could be fed. Pasteur replied to this: "You do not take account of the character of the water with which the alcohol is diluted. Like all common waters, even the purest, it contains ammoniacal salts and mineral matters that can feed the plant, as I have directly demonstrated. You have, moreover, not carefully examined the surface of the chips with the microscope. If you had, you would have seen the little articles of the Mycoderma aceti, sometimes joined into an extremely thin pellicle that may be lifted off. If you will send me some chips from the factory at Munich, selected by yourself in the presence of its director, I will, after drying them quickly in a stove, show the mycoderm on their surface to a committee of the Academy charged with the determination of this debate." Liebig did not accept the challenge, but the question involved has been decided.

The experiments in fermentation led by natural steps to the debate on spontaneous generation, in which Pasteur was destined to settle a question that had interested men ever since they lived. The theory that life originates spontaneously from dead matter had strong advocates, among the most earnest of whom was M. Pouchet. He made a very clear presentment of the question at issue, saying: "The adversaries of spontaneous generation assume that the germs of microscopic beings exist in the air and are carried by it to considerable distances. Well! what will they say if I succeed in producing a generation of organized beings after an artificial air has been substituted for that of the atmosphere?" Then he proceeded with an experiment in which all his materials and vessels seemed to have been cleansed of all germs that might possibly have existed in them. In eight days a mold appeared in the infusion, which had been put boiling-hot into the boiling-hot medium." Where did the mold come from," asked M. Pouchet, triumphantly, "if it was not spontaneously developed?" "Yes," said M. Pasteur, in the presence of an enthusiastic audience, for Paris had become greatly excited on the subject, "the experiment has been performed in an irreproachable manner as to all the points that have attracted the attention of the author; but I will show that there is one cause of error that M. Pouchet has not perceived, that he has not thought of, and no one else has thought of, which makes his experiment wholly illusory. He used mercury in his tub, without purifying it, and I will show that that was capable of collecting dust from the air and introducing it to his apparatus." Then he let a beam of light into the darkened room, and showed the air full of floating dust. He showed that the mercury had been exposed to atmospheric dust ever since it came from the mine, and was so impregnated and covered with it as to be liable to soil everything with which it came in contact. He instituted experiments similar to those of M. Pouchet, but with all the causes of error that had escaped him removed, and no life appeared. The debate, which continued through many months, and was diversified by a variety of experiments and counter-experiments, was marked by a number of dramatic passages and drew the attention of the world. M. Pasteur detected a flaw in every one of M. Pouchet's successful experiments, and followed each one with a more exact experiment of his own, which was a triumph for his position. Having shown, by means of bottles of air collected from different heights in a mountain-region, that the number of germs in the air diminishes with the elevation above the earth, and that air can be got free from germs and unproductive, M. Pasteur asserted decisively: "There is no circumstance now known that permits us to affirm that microscopic beings have come into the world without germs, without parents like themselves. Those who affirm it have been victims of illusions, of experiments badly made, and infected with errors which they have not been able to perceive or avoid. Spontaneous generation is a chimera." M. Flourens, Perpetual Secretary of the Academy, said: "The experiments are decisive. To have animalcules, what is necessary, if spontaneous generation is real? Air and putrescible liquids. Now, M. Pasteur brings air and putrescible liquids together, and nothing comes of it. Spontaneous generation, then, is not. To doubt still is not to comprehend the question." There were, however, some who still doubted, and to satisfy them M. Pasteur offered, as a final test, to show that it was possible to secure, at any point, a bottle of air containing no germs, which would, consequently, give no life. The Academy's committee approved the proposition; but M. Pouchet and his friends pleaded for delay, and finally retired from the contest.

The silk-raising industry of the south of France was threatened with ruin by a disease that was destroying the silk-worms, killing them in the egg, or at a later stage of growth. Eggs, free from the disease, were imported from other countries. The first brood flourished, but the next one usually fell victims to the infection, and the malady spread. All usual efforts to prevent it or detect its cause having failed, a commission was appointed to make special investigations, and M. Pasteur was asked to direct them personally. He did not wish to undertake the work, because it would withdraw him from his studies of the ferments. He, moreover, had never had anything to do with silk-worms. "So much the better," said Dumas. "You know nothing about the matter, and have no ideas to interfere with those which your observations will suggest." Theories were abundant, but the most recent and best authorities agreed that the diseased worms were beset by corpuscles, visible only under the microscope. He began his investigations with the idea that these corpuscles were connected with the disease, although assurances were not wanting that they also existed in a normal condition of the silk-worm. M. Pasteur's wife and daughters, and his assistants in the normal school, associated themselves with him in the studies, and became, for the time, amateur silk-raisers. He studied the worms in every condition, and the corpuscles in every relation, for five years. He found that there were two diseases—the contagious, deadly *pébrine*, the work of the corpuscles, and *flachery*, produced by an internal organism; and "became so well acquainted with the causes of the trouble and their different manifestations that he could, at will, give *pébrine* or *flachery*. He became able to graduate the intensity of the disease, and make it appear at any day and almost at any hour." He found the means of preventing the disorders, and "restored its wealth to the desolated silk district." The cost of this precious result was a paralysis of the left side, from which he has never fully recovered.

As early as 1860 M. Pasteur expressed the hope that he might "be able to pursue his investigations far enough to prepare the way for a more profound study of the origin of diseases." Reviewing, at the conclusion of his "Studies on Beer," the principles which had directed his labors for twenty years, he wrote that the etiology of contagious diseases was, perhaps, on the eve of receiving an unexpected light. Robert Boyle had said that thorough understanding of the nature of fermentations and ferments might give the key to the explanation of many morbid phenomena. The

German doctor, Traube, had in 1864 explained the ammoniacal fermentation of urine, by reference to Pasteur's theory. The English surgeon. Dr. Lister, wrote in 1874 to Pasteur that he owed to him the idea of the antiseptic treatment of wounds which he had been practicing since 1865. Professor Tyndall wrote to him, in 1876, after having read his investigations for the second time: "For the first time in the history of science we have a right to entertain the sure and certain hope that, as to epidemic diseases, medicine will shortly be delivered from empiricism and placed upon a really scientific basis. When that great day shall come, mankind will, in my opinion, recognize that it is to you that the greatest part of its gratitude is due."

The domestic animals of France and other countries had been subject to a carbuncular disease, like the malignant pustule of man, which took different forms and had different names in different species, but was evidently the same in nature. A medical commission had, between 1849 and 1852, made an investigation of it and found it transmissible by inoculation from animal to animal. Drs. Davaine and Rayer had, at the same time, found in the blood of the diseased animals minute filiform bodies, to which they paid no further attention for thirteen years, or till after Pasteur's observations on fermentation had been widely spread. Then, Davaine concluded that these corpuscles were the source of the disease. He was contradicted by MM. Jaillard and Leplat, who had inoculated various animals with matter procured from sheep and cows that had died of the disease without obtaining a development of the bodies in question. Davaine suggested that they had used the wrong matter, but they replied that they had obtained it direct from an unmistakable source. Their views were supported by the German Dr. Koch and M. Paul Bert. At this point, M. Pasteur stepped in and began experiments after methods which had served him as sure guides in his studies of twenty years. They were at once simple and delicate. "Did he wish, for example, to demonstrate that the microbe-ferment of the butyric fermentation was also the agent in decomposition? He would prepare an artificial liquid, consisting of phosphate of potash, magnesia, and sulphate of ammonia, added to the solution of fermentable matter, and in the medium thus formed would develop the microbe-ferment from a pure sowing of it. The microbe would multiply and provoke fermentation. From this liquid he would pass to a second and then to a third fermentable solution of the same composition, and so on, and would find the butyric fermentation appearing in each successively. This method had been sovereign in his studies since 1857. He now proposed to isolate the microbe of blood infected with carbuncle, cultivate it in a pure state, and study its action on animals." As he was still suffering from a partial paralysis, he employed M. Joubert to assist him and share his honors. In April, 1877, he claimed before the Academy of Sciences that he had demonstrated, beyond the possibility of a reply, that the bacillus discovered by Davaine and Rayer in 1850 was in fact the only agent in producing the disease. It still remained to reconcile the facts adduced by Messrs. Jaillard and Leplat with this assertion. The animals which they had inoculated died, but no bacteria could be found in them. M. Paul Bert, in similar experiments, had found a disease to persist after all bacteria had been destroyed. An explanation of the discrepancy was soon found.

The bacteria of carbuncle are destroyed as soon as putrefaction sets in. The virus with which these gentlemen had experimented was taken from animals that had been dead twenty-four hours and had begun to putrefy. They had inoculated with putrefaction, and produced septicæmia instead of carbuncle. All the steps in this line of argument were established by irrefragable proof. M. Pasteur afterward had a similar controversy with some physicians of Turin, at the end of which they shrank from the test experiment he offered to go and make before them. "Remember," shortly afterward said a member of the Academy of Sciences to a member of the Academy of Medicine, who was going—in a scientific sense—to "choke" M. Pasteur, "M. Pasteur is never mistaken."

Having discovered and cultivated the microbe that produces hen-cholera, Pasteur turned his attention to the inquiry whether it would be possible to apply a vaccination to the prevention of these terrible diseases of domestic animals. He found that he could transplant the microbe of hen-cholera to an artificially prepared medium and cultivate it there, and transplant it and cultivate it again and again, to the hundredth or even the thousandth time, and it would retain its full strength—provided too long an interval was not allowed to elapse between the successive transplantations and cultures. But if several days or weeks or months passed without a renewal of the medium, the culture being all the time exposed to the action of oxygen, the infection gradually lost in intensity. A virus was produced of a strength that would make sick, but not kill. Hens were inoculated with this, and then, after having recovered from its effects, with virus of full power. It made them sick, but they recovered. A preventive of hen-cholera had been found. In the experiments upon the feasibility of applying a similar remedy to carbuncular diseases, it was necessary to ascertain whether or not animals, which had once been stricken with the disease, were exempt from liability to a second attack. The investigator was met at once by the formidable difficulty that no animals were known to have recovered from a first attack, to serve as subjects for trial. A fortunate accident in the failure of another investigator's experiment gave M. Pasteur a few cows that had survived the disease. They were inoculated with virus of the strongest intensity, and were not affected. It was demonstrated, then, that the disease would not return. M. Pasteur now cultivated an attenuated carbuncle-virus, and, having satisfied himself that vaccination with it was effective, declared himself ready for a public test-experiment. Announcing his success to his friends, he exclaimed in patriotic self-forgetfulness, "I should never have been able to console myself, if such a discovery as I and my assistants have just made had not been a French discovery!"

Twenty-four sheep, a goat, and six cows were vaccinated, while twenty-five sheep and four cows were held in reserve, unvaccinated, for further experiment. After time had been given for the vaccination to produce its effect, all of the animals, sixty in number, were inoculated with undiluted virus. Forty-eight hours afterward, more than two hundred persons met in the pasture to witness the effect. Twenty-one of the unvaccinated sheep and the goat were dead, and two more of the sheep were dying, while the last one died the same evening; the unvaccinated cows were suffering severely from fever and œdema. The vaccinated sheep were all well and lively, and the vaccinated cows had neither tumor nor fever of any kind, and were feeding quietly. Vaccination is now employed regularly in French pastures; five hundred thousand cases of its application had been registered at the end of 1883; and the mortality from carbuncle has been reduced ten times.

There is no need to follow M. Pasteur in his further researches in the *rouget* of pork, in boils, in puerperal fever, in all of which, with other maladies, he has applied the same methods with the same exactness that have characterized all his work. His laboratory at the *École Normale* is a collection of animals to be experimented upon—mice, rabbits. Guinea-pigs, pigeons, and other suitable subjects, with the dogs upon which he is now studying hydrophobia most prominent. There is nothing cruel in his work. His inoculations are painless, except as the sickness they induce is a pain, and the suffering they cause is as nothing compared with that which they are destined to save. On this subject he himself has remarked in one of his lectures: "I could never have courage to kill a bird in hunting; but, in making experiments, I have no such scruples. Science has a right to invoke the sovereignty of the end."

What he has done, M. Pasteur regards as only the beginning of what is to be accomplished in the same line. "You will see," he has sometimes said, "how this will grow as it goes on. Oh, if I only had time!"

Pasteur's Researches in Germ-Life[2]

The weightiest events of life sometimes turn upon small hinges; and we now come to the incident which caused M. Pasteur to quit a line of research the abandonment of which he still regrets. A German manufacturer of chemicals had noticed that the impure commercial tartrate of lime, sullied with organic matters of various kinds, fermented on being dissolved in water and exposed to summer heat. Thus prompted, Pasteur prepared some pure, right-handed tartrate of ammonia, mixed with it albuminous matter, and found that the mixture fermented. His solution, limpid at first, became turbid, and the turbidity he found to be due to the multiplication of a microscopic organism, which found in the liquid its proper aliment.

---

[2] Pr John Tyndall

Pasteur recognized in this little organism a *living ferment.* This bold conclusion was doubtless strengthened, if not prompted, by the previous discovery of the yeast-plant—the alcoholic ferment—by Cagniard-Latour and Schwann.

Pasteur next permitted his little organism to take the carbon necessary for its growth from the pure paratartrate of ammonia. Owing to the opposition of its two classes of crystals, a solution of this salt, it will be remembered, does not turn the plane of polarized light either to the right or to the left. Soon after fermentation had set in, a rotation to the left was noticed, proving that the equilibrium previously existing between the two classes of crystals had ceased. The rotation reached a maximum, after which it was found that all the right-handed tartrate had disappeared from the liquid. The organism thus proved itself competent to select its own food.

It found, as it were, one of the tartrates more digestible than the other, and appropriated it, to the neglect of the other. No difference of chemical constitution determined its choice; for the elements, and the proportions of the elements, in the two tartrates were identical. But the peculiarity of structure which enabled the substance to rotate the plane of polarization to the right also rendered it a fit aliment for the organism. This most remarkable experiment was successfully made with the seeds of our common mold (*Penicillium glaucum*).

Here we find Pasteur unexpectedly landed amid the phenomena of fermentation. With true scientific instinct he closed with the conception that ferments are, in all cases, living things, and that the substances formerly regarded as ferments are in reality the food of the ferments. Touched by this wand, difficulties fell rapidly before him. He proved the ferment of lactic acid to be an organism of a certain kind. The ferment of butyric acid he proved to be an organism of a different kind. He was soon led to the fundamental conclusion that the capacity of an organism to act as a ferment depended on its power to live without air. The fermentation of beer was sufficient to suggest this idea. The yeast-plant, like many others, can live either with or without free air. It flourishes best in contact with free air, for it is then spared the labor of wresting from the malt the oxygen required for its sustenance. Supplied with free air, however, it practically ceases to be a ferment; while in the brewing-vat, where the work of fermentation is active, the budding *torula* is completely cut off by the sides of the vessel, and by a deep layer of carbonic-acid gas, from all contact with air. The butyric ferment not only lives without air, hut Pasteur showed that air is fatal to it. He finally divided microscopic organisms into two great classes, which he named respectively ærobies and anærobies, the former requiring free oxygen to maintain life, the latter capable of living without free oxygen, but able to wrest this element from its combinations with other elements. This destruction of pre-existing compounds and formation of new ones, caused by the increase and multiplication of the organism, constitute the process of fermentation.

Under this head are also rightly ranked the phenomena of putrefaction. As M. Radot well expresses it, the fermentation of sugar may be described as the putrefaction of sugar. In this particular field M. Pasteur, whose contributions to the subject are of the highest value, was preceded by Schwann, a man of great merit, of whom the world has heard too little. Schwann placed decoctions of meat in flasks, sterilized the decoctions by boiling, and then supplied them with calcined air, the power of which to support life he showed to be unimpaired. Under these circumstances putrefaction never set in. Hence the conclusion of Schwann, that putrefaction was not due to the contact of air, as affirmed by Gay-Lussac, but to something suspended in the air which heat was able to destroy. This something consists of living organisms which nourish themselves at the expense of the organic substance, and cause its putrefaction.

The grasp of Pasteur on this class of subjects was embracing. He studied acetic fermentation, and found it to be the work of a minute fungus, the *mycoderma aceti,* which, requiring free oxygen for its nutrition, overspreads the surface of the fermenting liquid. By the alcoholic ferment the sugar of the grape-juice is transformed into carbonic-acid gas and alcohol, the former exhaling, the latter remaining in the wine. By the *mycoderma aceti,* the wine is, in its turn, converted into vinegar. Of the experiments made in connection with this subject one deserves especial mention. It is that in which Pasteur suppressed all albuminous matters, and carried on the fermentation with purely crystallizable substances. He studied the deterioration of vinegar, revealed its cause, and the means of preventing it. He defined the part played by the little eel-like organisms which sometimes swarm in vinegar-casks, and ended by introducing important ameliorations and improvements in the manufacture of vinegar. The discussion with Liebig and other minor discussions of a similar nature, which M. Radot has somewhat strongly emphasized, I will not here dwell upon.

It was impossible for an inquirer like Pasteur to evade the question, Whence come these minute organisms which are demonstrably capable of producing effects on which vast industries are built and on which whole populations depend for occupation and sustenance? He thus found himself face to face with the question of spontaneous generation, to which the researches of Pouchet had just given fresh interest. Trained as Pasteur was in the experimental sciences, he had an immense advantage over Pouchet, whose culture was derived from the sciences of observation. One by one the statements and experiments of Pouchet were explained or overthrown, and the doctrine of spontaneous generation remained discredited until it was revived with ardor, ability, and, for a time, with success, by Dr. Bastian.

A remark of M. Radot's on page 103 needs some qualification. "The great interest of Pasteur's method consists," he says, "in its proving unanswerably that the origin of life in infusions which have been heated to the boiling-point is solely due to the solid particles suspended in the air." This means that living germs cannot exist *in* the liquid *when once raised to a temperature of 212° Fahr. No doubt* a great number of organisms collapse at this temperature; some, indeed, as M. Pasteur has shown, are destroyed at a temperature 90° below the boiling-point. But this is by no means universally the case. The spores of the hay-bacillus, for example, have in numerous instances successfully resisted the boiling temperature for one, two, three, four hours; while in one instance *eight hours'* continuous boiling failed to sterilize an infusion of desiccated hay. The knowledge of this fact caused me a little anxiety some years ago when a meeting was projected between M. Pasteur and Dr. Bastian. For though, in regard to the main question, I knew that the upholder of spontaneous generation could not win, on the particular issue touching the death temperature he might have come off victor.

The manufacture and maladies of wine next occupied Pasteur's attention. He had, in fact, got the key to this whole series of problems, and he knew how to use it. Each of the disorders of wine was traced to its specific organism, which, acting as a ferment, produced substances the reverse of agreeable to the palate. By the simplest of devices, Pasteur, at a stroke, abolished the causes of wine-disease. Fortunately the foreign organisms which, if unchecked, destroy the best red wines, are extremely sensitive to heat. A temperature of

50°C. (122° Fahr.) suffices to kill them. Bottled wines once raised to this temperature, for a single minute, are secured from subsequent deterioration. The wines suffer in no degree from exposure to this temperature. The manner in which Pasteur proved this, by invoking the judgment of the wine-tasters of Paris, is as amusing as it is interesting.

Moved by the entreaty of his master, the illustrious Dumas, Pasteur took up the investigation of the diseases of silk-worms at a time when the silk-husbandry of France was in a state of ruin. In doing so he did not, as might appear, entirely forsake his former line of research. Previous investigators had got so far as to discover vibratory corpuscles in the blood of the diseased worms, and with such corpuscles Pasteur had already made himself intimately acquainted. He was, therefore, to some extent at home in this new investigation.

The calamity was appalling, all the efforts made to stay the plague having proved futile. In June, 1865, Pasteur betook himself to the scene of the epidemic, and at once commenced his observations. On the evening of his arrival he had already discovered the corpuscles, and shown them to others. Acquainted as he was with the work of living ferments, bis mind was prepared to see in the corpuscles the cause of the epidemic. He followed them through all the phases of the insect's life—through the eggs, through the worm, through the chrysalis, through the moth. He proved that the germ of the malady might be present in the eggs and escape detection. In the worm, also, it might elude microscopic examination. But in the moth it reached a development so distinct as to render its recognition immediate. From healthy moths, healthy eggs were sure to spring; from healthy eggs, healthy worms; from healthy worms, fine cocoons; so that the problem of the restoration to France of its silk-husbandry reduced itself to the separation of the healthy from the unhealthy moths, the rejection of the latter, and the exclusive employment of the eggs of the former. M. Radot describes bow this is now done on the largest scale, with the most satisfactory results.

The bearing of this investigation on the parasitic theory of communicable diseases was thus illustrated: Worms were infected by permitting them to feed for a single meal on leaves over which corpusculous matter had been spread; they were infected by inoculation, and it was shown how they infected each other by the wounds and scratches of their own claws. By the association of healthy with diseased worms, the infection was communicated to the former. Infection at a distance was also produced by the wafting of the corpuscles through the air. The various modes in which communicable diseases are diffused among human populations were illustrated by Pasteur's treatment of the silk-worms. "It was no hypothetical infected medium—no problematical pythogenic gas that killed the worms. It was a definite organism." The disease thus far described is that called *pébrine,* which was the principal scourge at the time. Another formidable malady was also prevalent, called *flacherie,* the cause of which and the mode of dealing with it were also pointed out by Pasteur.

Overstrained by years of labor in this field, Pasteur was smitten with paralysis in October, 1868. But this calamity did not prevent him from making a journey to Alais in January, 1869, for the express purpose of combating the criticisms to which his labors bad been subjected. Pasteur is combustible, and contradiction readily stirs him into flame. No scientific man now living has fought so many battles as he. To enable him to render his experiments decisive, the French emperor placed a villa at his disposal near Trieste, where silk-worm culture bad been carried on for some time at a loss. The success here is described as marvelous; the sale of cocoons giving to the villa a net profit of twenty-six millions of francs. From the imperial villa M. Pasteur addressed to me a letter, a portion of which I have already published. It may perhaps prove usefully suggestive to our Indian or colonial authorities if I reproduce it here:

"Permettez-moi de terminer ces quelques lignes que je dois dicter, vaincu que je suis par la maladie, en vous faisant observer que vous rendriez service aux colonies de la Grande-Bretagne en répandant la connaissance de ce livre, et des principes que j'établis touchant la maladie des vers

à soie. Beaucoup de ces colonies pourraient cultiver le mûrier avec succès, et, en jetant les yeux sur mon ouvrage, vous vous convaincrez aisément qu'il est facile aujourd'hui, non seulement d'éloigner la maladie régnante, mais en outre de donner aux récoltes de la soie une prospérité qu'elles n'ont jamais eue."

The studies on wine prepare us for the "Studies on Beer," which followed the investigation of silk-worm diseases. The sourness, putridity, and other maladies of beer Pasteur traced to special "ferments of disease," of a totally different form, and therefore easily distinguished from the true *torula* or yeast-plant. Many mysteries of our breweries were cleared up by this inquiry. Without knowing the cause, the brewer not unfrequently incurred heavy losses through the use of bad yeast. Five minutes' examination with the microscope would have revealed to him the cause of the badness, and prevented him from using the yeast. He would have seen the true *torula* overpowered by foreign intruders. The microscope is, I believe, now everywhere in use. At Burton-on-Trent its aid was very soon invoked. At the conclusion of his studies on beer M. Pasteur came to London, where I had the pleasure of conversing with him. Crippled by paralysis, bowed down by the sufferings of France, and anxious about his family at a troubled and an uncertain time, he appeared low in health and depressed in spirits. His robust appearance when he visited London, on the occasion of the Edinburgh Anniversary, was in marked and pleasing contrast with my memory of bis aspect at the time to which I have referred.

While these researches were going on, the germ theory of infectious disease was noised abroad. The researches of Pasteur were frequently referred to as bearing upon the subject, though Pasteur himself kept clear for a long time of this special field of inquiry. He was not a physician, and he did not feel called upon to trench upon the physician's domain. And now I would beg of him to correct me if, at this point of the introduction, I should be betrayed into any statement that is not strictly correct.

In 1876 the eminent microscopist, Professor Cohn, of Breslau, was in London, and he then handed me a number of his "Beiträge, containing a memoir by Dr. Koch on splenic fever (*Milzbrand, Charbon,* malignant pustule), which seemed to me to mark an epoch in the history of this formidable disease. With admirable patience, skill, and penetration, Koch followed up the life-history of *bacillus anthracis,* the contagium of this fever. At the time here referred to he was a young physician, holding a small appointment in the neighborhood of Breslau, and it was easy to predict, as I predicted at the time, that he would soon find himself in a higher position. When I next heard of him he was head of the Imperial Sanitary Institute of Berlin. Koch's recent history is pretty well known in England, while his appreciation by the German Government is shown by the rewards and honors lately conferred upon him.

Koch was not the discoverer of the parasite of splenic fever. Davaine and Rayer, in 1850, had observed the little microscopic rods in the blood of animals which had died of splenic fever. But they were quite unconscious of the significance of their observation, and for thirteen years, as M. Radot informs us, strangely let the matter drop. In 1863 Davaine's attention was again directed to the subject by the researches of Pasteur, and he then pronounced the parasite to be the cause of the fever. He was opposed by some of his fellow-countrymen; long discussions followed, and a second period of thirteen years, ending with the publication of Koch's paper, elapsed, before M. Pasteur took up the question. I always, indeed, assumed that from the paper of the learned German came the impulse toward a line of inquiry in which M. Pasteur has achieved such splendid results. Things presenting themselves thus to my mind, M. Radot will, I trust, forgive me if say that it was with very great regret that I perused the disparaging references to Dr. Koch which occur in the chapter on splenic fever.

After Koch's investigation, no doubt could be entertained of the parasitic origin of this disease. It completely cleared up the perplexity previously existing as to the two forms—the one fugitive, the other permanent—in which the contagion presented itself. I may say that it was on the conversion of the permanent hardy form into the fugitive and sensitive one, in the case of *bacillus subtilis* and other organisms, that the method of sterilizing by "discontinuous heating" introduced by me in February, 1877, was founded. The difference between an organism and its spores, in point of durability, had not escaped the penetration of Pasteur. This difference Koch showed to be of paramount importance in splenic fever. He, moreover, proved that while mice and Guinea-pigs were infallibly killed by the parasite, birds were able to defy it.

And here we come upon what may be called a band-specimen of the genius of Pasteur, which strikingly illustrates its quality. Why should birds enjoy the immunity established by the experiments of Koch? Here is the answer. The temperature which prohibits the multiplication of *bacillus anthracis* in infusions is 44° C. (111° Fahr.). The temperature of the blood of birds is from 41° to 42° Fahr. It is therefore close to the prohibitory temperature. But then the blood globules of a living fowl are sure to offer a certain resistance to any attempt to deprive them of their oxygen—a resistance not experienced in an infusion. May not this resistance, added to the high temperature of the fowl, suffice to place it beyond the power of the parasite? Experiment alone could answer this question, and Pasteur made the experiment. By placing its feet in cold water he lowered the temperature of a fowl to 37° or 38° Fahr. He inoculated the fowl, thus chilled, with the splenic-fever parasite, and in twenty-four hours it was dead. The argument was clinched by inoculating a chilled fowl, permitting the fever to come to a head, and then removing the fowl, wrapped in cotton-wool, to a chamber with a temperature of 35° Fahr. The strength of the patient returned as the career of the parasite was brought to an end, and in a few hours health was restored. The sharpness of the reasoning here is only equaled by the conclusiveness of the experiment, which is full of suggestiveness as regards the treatment of fevers in man.

Pasteur had little difficulty in establishing the parasitic origin of fowl-cholera; indeed, the parasite had been observed by others before him. But, by his successive cultivations, he rendered the solution sure. His next step will remain forever memorable in the history of medicine. I allude to what he calls "virus attenuation." And here it may be well to throw out a few remarks in advance. When a tree, or a bundle of wheat or barley straw, is burned, a certain amount of mineral matter remains in the ashes — extremely small in comparison with the bulk of the tree or of the straw, but absolutely essential to its growth. In a soil lacking, or exhausted of, the necessary mineral constituents, the tree cannot live, the crop cannot grow. Now, contagia are living things, which demand certain elements of life just as inexorably as trees, or wheat, or barley; and it is not difficult to see that a crop of a given parasite may so far use up a constituent existing in small quantities in the body, but essential to the growth of the parasite, as to render the body unfit for the production of a second crop. The soil is exhausted, and, until the lost constituent is restored, the body is protected from any further attack of the same disorder. Such an explanation of non-recurrent diseases naturally presents itself to a thorough believer in the germ theory, and such was the solution which, in reply to a question, I ventured to offer nearly fifteen years ago to an eminent London physician. To exhaust a soil, however, a parasite less vigorous and destructive than the really virulent one may suffice; and, if, after having by means of a feebler organism exhausted the soil, without fatal result, the most highly virulent parasite be introduced into the system, it will prove powerless. This, in the language of the germ theory, is the whole secret of vaccination.

The general problem, of which Jenner's discovery was a particular case, has been grasped by Pasteur, in a manner, and with results, which five short years ago were simply unimaginable. How much "accident" had to do with shaping the course of his inquiries I know not. A mind like his resembles a photographic plate, which is ready to accept and develop luminous impressions, sought and unsought. In the chapter on fowl-cholera is described how Pasteur first obtained his attenuated virus. By successive cultivations of the parasite he showed that, after it had been a hundred times reproduced, it continued to be as virulent as at first. One necessary condition was, however, to be observed. It was essential that the cultures should rapidly succeed each other—that the organism, before its transference to a fresh cultivating liquid, should not be left long in contact with air. When exposed to air for a considerable time the virus becomes so enfeebled that when fowls are inoculated with it, though they sicken for a time, they do not die. But this "attenuated" virus, which M. Radot justly calls "benign," constitutes a sure protection against the virulent virus. It so exhausts the soil that the really fatal contagium fails to find there the elements necessary to its reproduction and multiplication.

Pasteur affirms that it is the oxygen of the air which, by lengthened contact, weakens the virus and converts it into a true vaccine. He has also weakened it by transmission through various animals. It was this form of attenuation that was brought into play in the case of Jenner.

The secret of attenuation had thus become an open one to Pasteur. He laid hold of the murderous virus of splenic fever, and succeeded in rendering it, not only harmless to life, but a sure protection against the virus in its most concentrated form. No man, in my opinion, can work at these subjects so rapidly as Pasteur without falling into errors of detail. But this may occur while his main position remains impregnable. Such a result, for example, as that obtained in presence of so many witnesses at Melun must surely remain an ever-memorable conquest of science. Having prepared his attenuated virus, and proved, by laboratory experiments, its efficacy as a protective vaccine, Pasteur accepted an invitation, from the President of the Society of Agriculture at Melun, to make a public experiment on what might be called an agricultural scale. This act of Pasteur's is, perhaps, the boldest thing recorded in this book. It naturally caused anxiety among his colleagues of the Academy, who feared that he had been rash in closing with the proposal of the president.

But the experiment was made. A flock of sheep was divided into two groups, the members of one group being all vaccinated with the attenuated virus, while those of the other group were left unvaccinated. A number of cows were also subjected to a precisely similar treatment. Fourteen days afterward all the sheep and all the cows, vaccinated and unvaccinated, were inoculated with a very virulent virus; and three days subsequently more than two hundred persons assembled to witness the result. The "shout of admiration," mentioned by M. Radot, was a natural outburst under the circumstances. Of twenty-five sheep which had not been protected by vaccination, twenty-one were already dead, and the remaining ones were dying. The twenty-five vaccinated sheep, on the contrary, were "in full health and gayety." In the unvaccinated cows intense fever was produced, while the prostration was so great that they were unable to eat. Tumors were also formed at the points of inoculation. In the vaccinated cows no tumors were formed; they exhibited no fever, nor even an elevation of temperature, while their power of feeding was unimpaired. No wonder that "breeders of cattle overwhelmed Pasteur with applications for vaccine." At the end of 1881 close upon thirty-four thousand animals had been vaccinated, while the number rose in 1883 to nearly five hundred thousand.

M. Pasteur is now exactly sixty-two years of age; but his energy is unabated. At the end of this volume we are informed that he has already taken up and examined with success, as far as his experiments have reached, the terrible and mysterious disease of rabies or hydrophobia. Those who hold all communicable diseases to be of parasitic origin, include, of course, rabies among the number of those produced and propagated by a living contagium. From his first contact with the disease Pasteur showed his accustomed penetration. If we see a man mad, we at once refer his madness to the state of his brain. It is somewhat singular that in the face of this fact the virus of a mad dog should be referred to the animal's saliva. The saliva is, no doubt, infected, but Pasteur soon proved the real seat and empire of the disorder to be the nervous system.

The parasite of rabies had not been securely isolated when M. Radot finished his task. But last May, at the instance of M. Pasteur, a commission was appointed, by the Minister of Public Instruction in France, to examine and report upon the results which he had up to that time obtained. A preliminary report, issued to appease public impatience, reached me before I quitted Switzerland this year. It inspires the sure and certain hope that, as regards the attenuation of the rabic virus, and the rendering of an animal, by inoculation, proof against attack, the success of M. Pasteur is assured. The commission, though hitherto extremely active, is far from the end of its labors; but the results obtained so far may be thus summed up:

Of six dogs unprotected by vaccination, three succumbed to the bites of a dog in a furious state of madness.

Of eight unvaccinated dogs, six succumbed to the intravenous inoculation of rabic matter.

Of five unvaccinated dogs, all succumbed to inoculation, by trepanning, of the brain.

Finally, of three-and-twenty vaccinated dogs, not one was attacked with the disease subsequent to inoculation with the most potent virus.

Surely results such as those recorded in this book are calculated, not only to arouse public interest, but public hope and wonder. Never before, during the long period of its history, did a day like the present dawn upon the science and art of medicine. Indeed, previous to the discoveries of recent times, medicine was not a science, but a collection of empirical rules dependent for their interpretation and application upon the sagacity of the physician. How does England stand in relation to the great work now going on around her? She is, and must be, behindhand. Scientific chauvinism is not beautiful in my eyes. Still, one can hardly see, without deprecation and protest, the English investigator handicapped in so great a race by short-sighted and mischievous legislation.

A great scientific theory has never been accepted without opposition. The theory of gravitation, the theory of undulation, the theory of evolution, the dynamical theory of heat—all had to push their way through conflict to victory. And so it has been with the germ theory of communicable diseases. Some outlying members of the medical profession dispute it still. I am told they even dispute the communicability of cholera. Such must always be the course of things, as long as men are endowed with different degrees of insight. Where the mind of genius discerns the distant truth, which it pursues, the mind not so gifted often discerns nothing but the extravagance, which it avoids. Names, not yet forgotten, could be given to illustrate these two classes of minds. As representative of the first class, I would name a man whom I have often named before, who, basing himself in great part on the researches of Pasteur, fought, in England, the battle of the germ theory with persistent valor, but whose labors broke him down before he saw the triumph which he *foresaw* completed. Many of my medical friends will understand that I allude here to the late Dr. William Budd, of Bristol.

The task expected of me is now accomplished, and the reader is here presented with a record in which the verities of science are endowed with the interest of romance.

Printed in Great Britain
by Amazon